How To Raise Strong & Healthy Chickens

I0482295

HTeBooks

Disclaimer

This book is designed to provide condensed information. It is not intended to reprint all the information that is otherwise available, but instead to complement, amplify and supplement other texts. You are urged to read all the available material, learn as much as possible and tailor the information to your individual needs.

Every effort has been made to make this book as complete and as accurate as possible. However, there may be mistakes, both typographical and in content. Therefore, this text should be used only as a general guide and not as the ultimate source of information. The purpose of this book is to educate.

The author or the publisher shall have neither liability nor responsibility to any person or entity regarding any loss or damage caused, or alleged to have been caused, directly or indirectly, by the information contained in this book.

Table of Contents

How Will This Book Help You?

With the current economic times coupled with the increasing number of diet related health problems, the best you can do is to have your own supply of your favorite foods. Whether it is growing fruits and veggies or raising some animals like poultry, you will definitely be a lot better placed than counting on the GMO chicken that you buy from the local meat market.

If you are thinking that it is impossible to rear chicken even in the tiniest of places, then think again. Rearing chicken is a lot easier than you have been thinking if you've never tried it. If you don't know what to do, then this book will offer all the help you need. It will help you to decide whether to keep layers or broilers, prepare the chicken coop, bring the chicks home, feed them and protect them from environmental hazards and diseases and feed them until they are fully grown and ready to lay eggs or ready for slaughter.

It doesn't matter which stage you are in rearing chicken. Whether you are just contemplating about buying the chicks or are thinking of how to deal with the diseases that affect the chicken in their growth phase, you will find this book helpful.

Beginning At The Beginning: Choosing The Type of Chicken to Raise

"Beginnings often spell the end. Thus, handle these as if they were the most important things in all of the earth"

-L. Ron Hubbard.

Rearing chicken is a process. The first thing you need to do is to decide what kind of chicken you want to keep; is it broilers or layers. So, how can you choose between these?

It is important to begin at the beginning: an egg laying poultry is referred to as an egger or layer whereas the broiler poultry is one reared primarily for meat. Thus, a layer should be raised to produce many large eggs without necessarily growing too much. On the other hand, you will need to raise a broiler to yield a lot of meat and hence, be able to grow well.

Good management practices will be essential for poultry production, regardless of whether you choose to go the broiler way or the layer way. The thing to understand here is that the management procedures differ for both layers and broilers. These will include temperature maintenance as well as hygienic conditions in both housing and poultry feed as well as keeping pests and diseases at bay. Let's take a quick overview of how these two are different in the way you raise them.

Broiler chicken

The nutritional, housing, and environmental requirements of broilers differ from those of layers. Broiler feed should be rich in vitamins to boost the growth rate and superior feed efficiency. The ration for broilers will also be rich in proteins as well as have adequate fat. Absolute care is to be taken to minimize mortality as well as maintain feathering as well as the quality of the carcass.

Layer chicken

These show 2 distinct phases in their life; the growing period and the laying period. During the growing period, the issue of space is vital: they need enough of it. If you overcrowd them, the result is suppression in growth. However when it comes to feed, it should be administered in a restricted and calculated manner.

When the laying period comes along, both adequate space and ample lighting are of great importance. Feed your chicken with minerals, vitamins, and micronutrients to influence the laying of eggs. You may use agricultural by-products, which will be cheaper yet more fibrous and beneficial as a result.

So, if you've settled for broiler or layers, you should start preparing where you will house them and provide ideal conditions for them to guarantee maximum growth and maximum productivity while ensuring minimal mortality.

Key Point

You need to settle on what it is you want most. Is it meat or is it eggs? You can as well settle for some mixed breeds that can supply eggs and meat for your family. But these will probably have different rates of growth; their growth is likely to be slower than that of the pure breeds.

Preparing A Home For Your Chicks

"Give me six hours to chop down a tree and I will spend the first

four sharpening the axe."

—Abraham Lincoln

Step I: What to do before the chicks arrive

Usually, this is mostly about setting up the brooder. It will make little sense to bring your load of chicks over and not have a home ready for them. This is a strong prerequisite to raising strong, healthy chicken. The brooder will need to provide adequate ventilation, protection, and warmth.

What to do

Make sure there is a ¾ square foot space per chick

Cover the brooder floor with about 4 inches of litter

Place a heat lamp above the brooder. For the 1st week, you will need to keep brooder temperatures at 95 degrees.

Getting your chicks settled

Regardless of where the chicks are from (feed mill, hatchery etc.), you need to put your chicks in a brooder right away.

Show your chicks the location of their water. You will do this by gently dipping their beaks in the water one by one.

Watch the chicks for a while to make sure they are neither too cold nor too hot.

Tip: Pasty butt: This occurs when poop is stuck on the chick's downy feathers, resulting in demise, as the chick cannot poop. When this happens, clean the refuse off with warm water and put a little bit of olive oil all around the vent.

Step II: Raising the broiler and layer chicks to strong and healthy chicken

This book emphasizes on the term "raise". Here is the thing: to raise, especially with regard to chicken, means to take care of them right from the time when they are tiny chicks with tiny pairs of legs. If you do the starting duties right, you increase the chances of having healthy chicken immeasurably. Here is what to do to ensure your chicks grow up to be healthy, strong chicken:

Brooder

The first thing to do is to confine the chicken to a brooder that has solid sides each about 18 inches high so as to keep drafts out. It is important to make sure that the brooder is located near a heat source, most preferably a heat lamp. For each chick, allocate about

6 square inches of floor space. Finally, the chicken need to be as protected from predators as possible. Locate the brooder in a safe and secure place.

Brooder Floor

Pine shavings are the best material you can use to cover the floor with. If pine shavings are a bit rare where you come from, any other absorbent bedding material will do the job. Steer clear from using kitty litter or cedar shavings. Stay away from using newspaper as well. For the duration of the first two days, cover the littler with several paper towels/tissue paper or even a piece of aged cloth so as to keep the chicks from eating the litter until they can locate their food. After this period however, you can remove the paper towels or pieces of cloth.

Temperature

Ensure to turn on the heat lamps at a day or more before you can bring the chicks home. This should make sure that the floor, walls, air, and bedding start warming up and even any disinfectant you've used has dried. You will also ensure that the air temperature is ideal.

For the 1st week, the chicks require a steady temperature of about 95 degrees Fahrenheit. This should be kept steady at all times. For every week that passes, drop the temperature by about 5 degrees until the surrounding room temperatures outside the brooder is equal to the temperature inside the brooder. However, drop the temperature no more once you hit the 60 degrees Fahrenheit mark.

Feed

For all chicks ensure to use starter- we will talk about this in detail later in the book. For the 1st two days, sprinkle the feed on a white paper plate or even some paper towels of the same color to as to make it easier to find. You should make sure to have feed available at all times in the feed dishes that you allocate the chicks.

Water

With baby chicks, you need to put their water in a container that is both shallow and narrow. The reason for this is to cut off any chances the chicks may have of drowning. Dip their beaks into the water containers gently as you place them into their brooder so that they may develop a memory of where it is. At all times, always have water available.

Note: Ensure to fill the chick's waterers with warm water then dissolve a tablespoon of molasses to boiling water to let it dissolve then add in some more water to make up to 4 liters or a gallon of lukewarm water. Do this before your chicks arrive. This should be the first thing you give to the chicks when they arrive (give them for 3 days) to facilitate recovery from the traveling. The chicks can survive for about 3 days by relying on the yoke sack that they draw into their body right before they are able to hatch but they need to drink water- use the strategy mentioned above to give them water.

Handling

It is a mistake to handle baby chicks too much. The reason for this is that handling them to a high degree only serves to stress them and in turn, makes them not grow well. Handling them too much may also kill the chicken eventually.

Key point

When chicks are contented, they usually are fairly quiet, often spread out over the brooder and either eating, drinking or eating. If you realize the chicken are making way too much noise, there is a high chance that something is wrong. Usually, the reason is that the chicks are too cold. If you notice them spread out against the walls of the brooder and panting heavily, then they are too hot for comfort.

Housing: The Chicken Coop & Chicken Run

"Once your mind is made up that chicken are for you, it is only logical that the very next thought to cross your mind should have everything to do with housing."

-Tim Daniels.

The Chicken Coop

As far as expenses go, nothing will be as expensive as a chicken coop for your chicken. But if your chicken are to truly grow up as healthy as can be, you really do have to invest in a good chicken coop. The good part about all this is that there are multiple designs and types out there in the market. As such, you will be able to come out of it feeling like you indeed have gotten value for your money.

Usually, it is a case of "you get what you shell out for" with wood but then again, there are lots of farmers who have been able to keep their cheap wooden chicken coops in superb conditions for periods of up to 7 years by simply employing the services of a regular coat of wood preserver. Regular maintenance (for instance, replacing any latches that have rusted and roof felts that the winds have picked up) does not hurt. Once you make up your mind on raising chicken and set out looking for a coop, then do take a very keen look on the overall workmanship and the quality/thickness of the wood used as this will be the number one indicator on whether you should be paying a higher price or not. Very obviously, the better the wood is, the longer your chicken coop will last. However, the wood will also be a lot more expensive.

Chicken houses, like most other things in life, can be either immensely beautiful or very basic. If anything, you could set out with a hammer and nails and build your own coop out of cheap pallet wood; as long as it does its job (which is to keep the inner conditions dry and draft free), everything else matters little.

Wood for your chicken coop, ideally, should always be moisture treated. This is to keep off the rot in the first year. The screws, fittings, and nails need to be galvanized to stave off the rust.

Here are a few things you must do to ensure the very best coop for the healthiest of chicken:

When you first buy a chicken coop, apply a coat of paint to give it a longer lease of life.

If you are treating your chicken coop, check to see if the product you are using is animal friendly. Leave it out to dry before you allow your birds into the coop. If you are unsure, call the numbers provided on the back of the tin and ask for advice.

Essential features of the house are nest boxes, ventilation (which should be adjustable) and roosting perches. However, you must consider the ease of collecting eggs as well as the ease of cleaning.

The other thing to keep in mind is that spending that little bit more on a chicken house that comes with a droppings board will save you hours of struggling in the long run.

The bare minimum, when it comes to chicken, is to have dry conditions, away from draughts, safe from any predatory animals and have a private place to lay eggs. Ask yourself if your coop provides all the above. If it doesn't, get back to the drawing board.

The Chicken Run

To remain healthy, your chicken will need to step outside. Your chicken run could be in the back garden or perhaps a pen or a small fenced area. Whatever it is that you settle on, the advice here is to have your run as big as possible so that your chicken can have as much free ranging space as is possible. There are multiple benefits to chicken in set ups that offer free ranging conditions, among them happier, healthier chicken that tend to cost you less to raise and even produce more eggs.

Keep the chicken run secure

The run should successfully keep the fox out and the chicken in. If it is possible, purchase wire netting that is at least 20cm into the ground to keep off predators that might try to dig under the fence. Cover the roof of the chicken run as well.

Where foxes and other chicken predators are a serious problem and the run in question is particularly large, you may employ the services of an electric fence. The ideal set up will have 3 strands of electric wire looped back and forth in an organized manner. The first strand, kept at low level, will stop the foxes from digging underneath the run. The other 2 strands will successfully keep out the predator from climbing over the fence.

Key point

For ideal conditions, keep the floor covered with wood shavings, straw or chopped cardboard. If anything, this material will get into your veggies patch far faster than anything else. The larger the pen, the easier it will be to infiltrate. Firm up security measures the more your pen size grows.

Feeding Ideals For Strong And Healthy Layers: Ideals On What, How And When To Feed Your Chicken At Different Ages And Stages As Well As What Their Feed Entails

"The spirit cannot endure the body when overfed, but, if underfed, the body cannot endure the spirit."

-St Frances de Sales

Just as the case is with human beings, chicken at differing stages of development will require different formulations of feed. The commercially prepared rations pack quite a punch in as much as offering a balanced diet is involved. This is the leading reason why you should strongly consider commercially processed feed. The other thing to keep in mind is that dabbling in assembling feed that is home processed is not recommended. If the calculations you put in place are imprecise, it is very easy to affect the growth in young chicken as well as egg production in older chicken.

Starter Feed (Chicks: Day 1 to 8 weeks)

Chicks that are day old through to 8 weeks will require starter feed that will contain 20% protein. You should know that starter feed will contain the highest amount of protein, going by percentages; a layer will ever take in its life. All this makes tremendous sense however, given the astronomical growth rate in the first couple of months of life.

You can have starter feed in two varieties: medicated and unmedicated varieties. The medicated feed will contain amprolium (which will crop up in the later chapters dealing with chicken diseases) that plays the role of keeping away illnesses like coccidiosis. If your chicks have been vaccinated, keep them away from medicated feed, as the amprolium will combat the vaccine rendering it useless. Medicated feed is unnecessary too if the living conditions of the chicks are kept clean to a high standard.

With regard to treats, you may stop obsessing right this instant on what age is appropriate to start including these in the diet. No age is ever appropriate, as treats take up a large nutritional part that would otherwise have been offered by normal feed. Treats must also be served along with grit, which is very useful for digestive purposes.

Grower Feed ("Teenage generation: 8-18 weeks of age)

With the vast amount of proteins it contains, starter feed may well launch the body of a young pullet into egg laying before it is even ready for it. This is where grower ration comes in: with its 16-18% protein content, it is significantly less compared to starter feed.

Keep off from feeding your chicks with layer feed until they are old enough for it (roughly 18 weeks) or have begun laying eggs. This is because it comes with calcium, which may well damage the chicken' kidneys, lead to kidney stones and greatly shorten the chicken' lifespan.

If you are providing treats along with grower feed to your chicken, always accompany grit with it, especially if the chicken you are raising are not allowed to wander around and forage.

Layer Feed (The big lot: 18 weeks and above)

Layer feed comes available in mash, pellet or crumble forms. Mash is the tiniest form while pellets are the largest, with crumble somewhere in between.

Layer feed comes with around 16 to 18 percent of protein along with additional calcium, necessary for the formation of eggshells. Hens that are already laying may be fed with layer ration as early as 18 weeks or even as late as it takes for them to produce their 1st egg. If the bird is younger than 18 weeks, it is wise to keep off the feed.

While layer feed will contain calcium, a source of additional calcium, such as oyster that is crushed ought to be made available to hens that are laying in a dish that is separate. Keep this away from the feed at all costs. Laying hens have varying needs of calcium and will take in as much as they need. Never add oyster shells directly to the feed, since the excess calcium may well be harmful to your chicken.

Cut the treats away

Commercial layer feed will provide every daily nutritional element that a chicken requires. Providing the chicken with snacks, table scraps and treats in addition to the regular feed will only serve to interfere with the chicken balanced diet, at least to a degree. Limiting the snacks, even those that may be categorized as healthy (homemade flock block, mealworms and pumpkin seeds) will ensure that your flock is getting all it needs. This will also help them swerve away from issues of obesity, egg binding, feather picking and a marked reduction in egg production.

Scratch

Lumping chicken scratch in with the feed category is misguided. Chicken scratch does not qualify as feed. Scratch content will vary from region to region but generally, it consists of cracked corn and an assortment of other grains. Scratch acts as a source of energy (thinking about carbs will help you understand this). However, scratch is a poor source of vitamins, proteins, and minerals. Feed your chicken scratch in a sparing form.

In cold weather, you will expect your chicken to expend more energy than is usual. Adding in a good amount of scratch just before dusk will be great in aiding them supplement energy. Too much scratch will bring about obesity as well as obesity related fatalities. The extra calories that come with scratch will also be beneficial to chicken that are brooding.

Feeding different age groups together

Chicken of differing age groups often occupy a similar living space at any given time. This raises the question of just how to feed each of them adequately. While this situation is hardly ideal, it isn't that unique either to be honest. Often, the best solution when dealing with a group of mixed flock is to feed them un-medicated starter or grower feed with calcium available in a separate dish. The additional protein in the starter or grower feed will not hurt those birds that are older but the calcium that is contained in layer feed may do damage to the kidneys of those birds that are growing.

Key Point

You should provide age appropriate food to your chicken otherwise you risk hurting them or even disrupting their productivity as you try to help. This is especially because chicken (layers) of different age groups will require nutrients in different amounts at different ages. For instance, while they may require more protein during the first 8 weeks to facilitate growth, they will need more calcium and other nutrients when laying.

Feeding Ideals For Strong And Healthy Broilers: Ideals On What, How And When To Feed Your Chicken At Different Ages And Stages As Well As What Their Feed Entails

"Tell me what you eat, and I will tell you who you are."

-Brillat-Savarin

Poultry feeds are usually rich in minerals, vitamins, energy, proteins, and any nutrients that the chicken might need for the health of the birds, egg production, and proper growth. This means that when you add in other food whether by substituting or combination, you are in essence upsetting the balance of nutrients in the food. This means feeding any supplement or grain isn't recommended.

Note: Broiler feeding plan can sometimes vary depending on the breed and your style. You can feed them with three different kinds of feed at different times.

Immediately the chicks hatch, feed them a "starter" until they reach about 6 to 8 weeks. The starter is usually high in protein making it very effective for fueling growth. But as the chick matures, it will start needing lesser percentage of dietary protein and start needing higher level of energy.

Note: If you are going for the 3-stage process, feed the chicken the "starter" when they are 1 day to about 3 weeks old. After that, start introducing the "grower" starting from week 4 then feed that until the chicken are about 6 weeks old.

When the chicks are about 6-8 weeks, you should ensure to feed them "finisher" diet or what is referred to as "developer" diet (to pullets or cockerels, which are kept for breeding purposes). You should feed them the finisher diet until they are ready for slaughter. You should feed cockerels and pullets the developer until they get to about 20 weeks.

The finisher is usually lower in protein and very high in energy. You can use it to enhance the fat content in the meat so if you want your broilers to have plenty of fat, ensure to use the finisher. However, if you want a leaner chicken, you can continue feeding them with the grower from week 3 until they are ready for slaughter.

Note: Chicken that are kept for egg production are usually fed on pullet-type diets and not broiler diets even if they are from egg-type or broiler type stock.

As you transition from one feed to another, you must ensure to introduce it slowly. To do this, ensure to increase the amount of feed in a particular feeder over a period of about a week.

Key point

Feed broilers on Starter between 1 day and 3 weeks, Grower between 3 and 6 weeks and finisher between 6 weeks and slaughter. Finisher is specially designed to supply energy and increase the amount of fat on the chicken so you can avoid it if you don't want too much fat.

Safety & Health: Focus On The Most Common Yet Lethal Of Illnesses, Their Symptoms, Prevention And Their Treatment

"They say that prevention is better than cure, which makes tremendous sense. At times though, you will have to use cure as prevention for something worse; something like death or stunted features. Thus in essence, both prevention and cure are the same thing albeit with differing trajectories of application"

-Anonymous

As far as health goes, it is obvious that your chicken will get nowhere near it if diseases consistently plague them. It is remarkable just how poultry and humans are in their life patterns when it comes to diseases. Some diseases are more common than others; some are more lethal than the rest of them. Finally, there are those that are both common and lethal at the same time; a terrible marriage of traits that is detrimental to chicken. This chapter focuses on the latter kind and how to boost the chances of your chicken being both strong and healthy by both putting in prevention measures and treating them when those measures are breached.

Vitamin A deficiency

Vitamin A, by default, is required for the well-being of chicken as well as the proper function of their mucous producing glands.

Deficiency of Vitamin A usually is because of a lack of vitamins in the diet/feed.

Symptoms: Crusty material is observed in both the eyelids and the nostrils. If it is left untreated, the accumulation of cheesy material is observed. It will often mimic respiratory diseases in its initial stages and your chicken, due to the similar damage it brings along with it, will have a lot of trouble swallowing. A severe drop in both egg production and weight is observed.

Prevention: Free-range chicken will get their fare share of Vitamin A from feeding on leafy greens. This is not the case with enclosed chicken, which will require it in their feed. Adding in a few leafy greens to the diet if you are rearing chicken in an enclosure will certainly not hurt.

Treatment: This one is as basic and direct as remedies come. Change the chicken feed and proceed to supplement it with 2 to 4 times the normal level for a duration of a fortnight. You may use a water-soluble supplement, which is available with considerable ease.

Coccidiosis

Symptoms: Your chicken will exhibit considerable weight loss, relentless bouts of diarrhea, and a loss in pigmentation. When the infection is allowed to sit so that it becomes quite severe, blood will be observed in the diarrhea and if attention is not given, fatality is the next stage.

Prevention: The way to prevent this terrible illness is absolutely reliant on drainage systems. Improve the drainage and rotate both the pen and water locations to cut down on the risk of infection in your chicken. Changing the topsoil yearly in a floor pen will reduce the risk of coccidiosis by eliminating the buildup of oocysts if any exists. You may also prevent the disease by using medicated starter as well as growth feeds.

Treatment: You will want to keep a solid batch of sulfa drugs and amprolium close by in case the disease strikes (which it does with alarming consistency). Administer these drugs in the drinking water of your chicken.

Mycoplasmosis

Symptoms: Your chicken will exhibit dirty nostrils, watery eyes, sneezing and bouts of coughing. Your chicken will also be slow in developing. Hatchability, fertility and production of eggs all register drops. Over time, cheesy material collects in both the sinuses and the eyelids. Noticeable outward swelling is also observed.

Prevention: You will find that the best prevention for MG is to purchase chicken that are free of the disease, if at all you get around to buying grown chicken rather than raising them from the chick stage. However, this is often more easily said than done given that the carriers often appear in perfect health. A basic and cheap blood test that is carried out by the bulk of veterinary diagnostic labs will detect any prior exposure to the disease. Start with contacting your veterinarian or local extension office and ask for an expert to help you.

Treatment: You will have to begin at the beginning here- lower any stress boosters that may be present. Clean the coops and reduce dust. Follow this up promptly with antibiotic treatment and proper nutrition. Both tetracycline and tylosin are some powerful antibiotics that will help do away with symptoms of the disease. However, if you are dealing with a carrier chicken, be warned that the antibiotics may not be able to do much as far as complete curing is concerned.

Administer antibiotics via the drinking water but be keen not to administer them for a period longer than 7 days. Vaccines are generally not encouraged in most states in the US, especially given that they tend to foster a milder version of the same disease they are supposed to help keep at bay.

Colibacillosis

Symptoms: Your chicken will appear disturbed and listless, show labored breathing and have their feathers almost permanently ruffled. They will also cough relentlessly. If the infection has been allowed to stew, your chicken will diarrhea, exhibit swelling as well as spleen and liver congestion. In the case of newly hatched chicken, navel infection may be observed.

Prevention: To cut down on the risk of infection, keep the area where your chicken live free of dust, feces and also ensure that there is sufficient living space for each of them. Do not use eggs that are visibly dirty for hatching purposes as these may well perpetuate the disease. Efforts to clean the eggs may lead to the eggshells cracking and this will open them up for increased bacterial penetration.

Treatment: The disease responds to antibiotic treatment. Especially where tetracycline and sulfa drugs are used, results are often positive. Make sure treatment exceeds 5 days before you evaluate any improvement in disease symptoms.

Key Point

Lowering stress conditions pops up in just about every preventative measure. Keep the dust away and the feces removed; make sure the drainage system is flawless and you will have made a huge step towards a disease-free chicken farm.

Different Chicken Breeds: What To Know About Varied Breeds Before Selection So As To Boost Your Chances Of Owning Strong And Healthy Chicken

"Selection is everything. By default, all the survivors of nature-that lot that has managed to outlive the rest and establish a line of offspring, were selected by mother earth to perpetuate the course. You too must learn to choose what to go for to truly succeed where you may."

-James May.

Here are some factual comments of different chicken breeds. While selection will ultimately come down to you, it will be a sapient thing to follow the recommendations given, where they appear.

Golden sex link

This breed of chicken is characterized by weak immune systems. This said, you should know that they are not the hardiest of breeds and are easily bullied by other chicken breeds. While you may perhaps have a different opinion on these, it is generally not recommended that you settle for these.

Source: Calranch.com

The Americana Breed

These are remarkably beautiful show birds. They are strange and moody though. These birds, while eye catching in their splendor, will not fit the profile of the utilitarian bird most people prefer. Especially, they will not fit the profile of birds that you would prefer to have on a working farm for the production of food.

Source: Imgkid.com

The leghorn

This breed is a phenomenal layer of eggs. If there is any fault to them, it is that they are generally distrustful of human folk. You are guaranteed to have some trouble catching these but when all is said and done, the eggs they lay will probably do a lot more for you than rewarding you for a tiring chase.

Source: Chickenbreedslist.com

The Delaware white breed

This bird is a great hybrid bird when it comes to producing eggs and providing the family with meat. However, in as much as they make a superb hybrid for both egg production and meat provision, they are not the absolute best for each one. Compared to other breeds, they grow to quite a large size and with this growth and girth, comes a remarkable appetite.

While their pure white appearance is really quite beautiful to behold, it ranks up there, along with the rest of the drawbacks package. Their pure white coat makes them easy prey for predators, as they are easy to spot from long distances. Birds that are more natural colored tend to blend in better with the natural terrain.

The barred rock breed

These birds are a favorite for many chicken farmers. It is not unnatural to hear a farmer reply to a barred rock-question with, "I like these birds. I just do". The reason for the love is that these birds are among the easiest to handle, are fairly productive and above all, are quite a hardy breed. This breed has been described as one of the most consistent picks of favorite breeds for farmers.

Source: imgbuddy.com

The Rhode Island Red

This breed is simply fantastic. It is, by a mile, the most potent survivor of all the breeds mentioned above. In the opinion of most experienced farmers in different parts of the country, no other breed comes as close to perfection as the Rhode Island Red, especially in a farm context. They are very friendly as well and are handled with ease. Here is a testimonial from one of the most renowned poultry bloggers in the country, "During the seasons of spring and summer, my Rhode Island flock will lay between 6 and 7 eggs per week, each."

This is quite remarkable in itself. Simply put, if you are a novice when it comes to chicken and are looking for a breed to kick things off with, just go with the Rhode Island Red breed. In addition to a supreme hardiness, friendliness, and easy handling, they are some of the most prolific layers around.

Source: Thechickentreet.com

Key Point

Hardiness, in addition to solid egg laying and meat production, is a key trait to look out for. Try to settle for breeds that show a tenacious will to survive.

How to Apply What You've Learned?

Perhaps, the best thing about this book is that while it does detail and caution against problems, it does not make out the practice of raising the "fair fowl" as nuclear science. The straightforward directives, instructions, and explanations will make it all the easier to apply what you learn.

Following what it details to the letter and succeeding while you try will require the support of an expert, at least in stretches. Call up a vet, perhaps one with whom you have enjoyed a professional relationship for some time, and ask as much info from them as possible. Above all, keep your chicken coop clean: this book will hardly help you if you do not keep the living places clean. If you have noticed, your success towards raising strong healthy chicken will depend on how clean you keep the coop.